fire

Quinc[illegible] Dept Store

KEEP BACK 300'

A Pictorial Report on Firefighting
Robert F. Stella
Quinlan Press

Hanover, General Alarm, Factory

Published by Quinlan Press
131 Beverly Street
Boston, MA 02114
(617) 227-4870
1-800-551-2500

Library of Congress
Catalog Card Number 86-61645
ISBN 0-933341-58-X

Printed in the United States of America
First printing March, 1987
Second printing October, 1987

Pembroke, General Alarm, Lumber Yard

To my mother and father for all of their love and encouragement.

Boston, 9 Alarms, Factory

I would like to thank all those who have helped in the making of this book and, in particular Captain Mark Duff, Captain William Antoine, Firefighter Mike Antoine and Donald Schmidt.

Quincy, 4

Bob Stella has been a freelance Photojournalist since 1975. His work has appeared in **LIFE Magazine, U.S. News & World Report, Firehouse Magazine, The National Fire Protection Association, Boston Globe** and **Boston Herald** as well as several other newspapers and magazines throughout the country. Bob is a member of the Boston Press Photographers Association, National Press Photographers Association and The Bare Cove Fire Museum, Hingham, Mass.

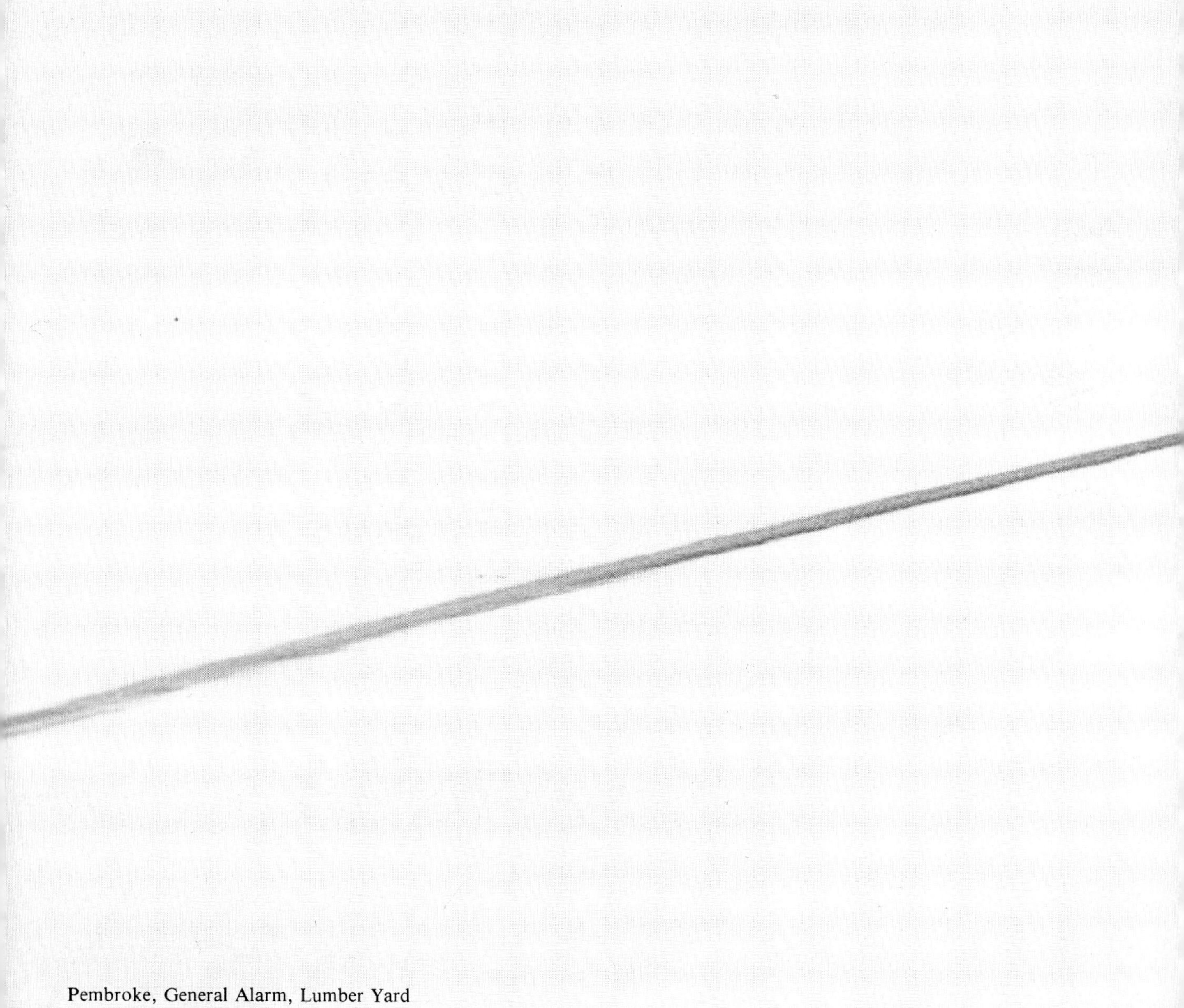

Pembroke, General Alarm, Lumber Yard

All photographs in this book were taken with Nikon F2 and FM Cameras with Nikon lenses ranging from 18mm to 300mm. Tri-x (ASA 400) black & white film was used sometimes pushed as high as ASA 6400. Color film is Ektachrome 400.

Weymouth, 3 Alarms, Navy Base

Quincy, Truck Fire

Hingham

Quincy, 2nd Alarm, House

Quincy, 2nd Alarm, Bank Under Construction

Quincy, 4th Alarm, Block of Stores

Braintree, Garage

Randolph, Victims of House Explosion

Brockton, 2nd Alarm, House

Quincy, 2nd Alarm, Injured Firefighter

Boston, 3rd Alarm

Quincy, 2nd Alarm, House

Quincy

Randolph, 2nd Alarm, House

Brockton

Boston, 9 Alarms

Quincy, 2nd Alarm, House

Braintree, 3 Alarms

Quincy, 3 Alarms, Block of Stores

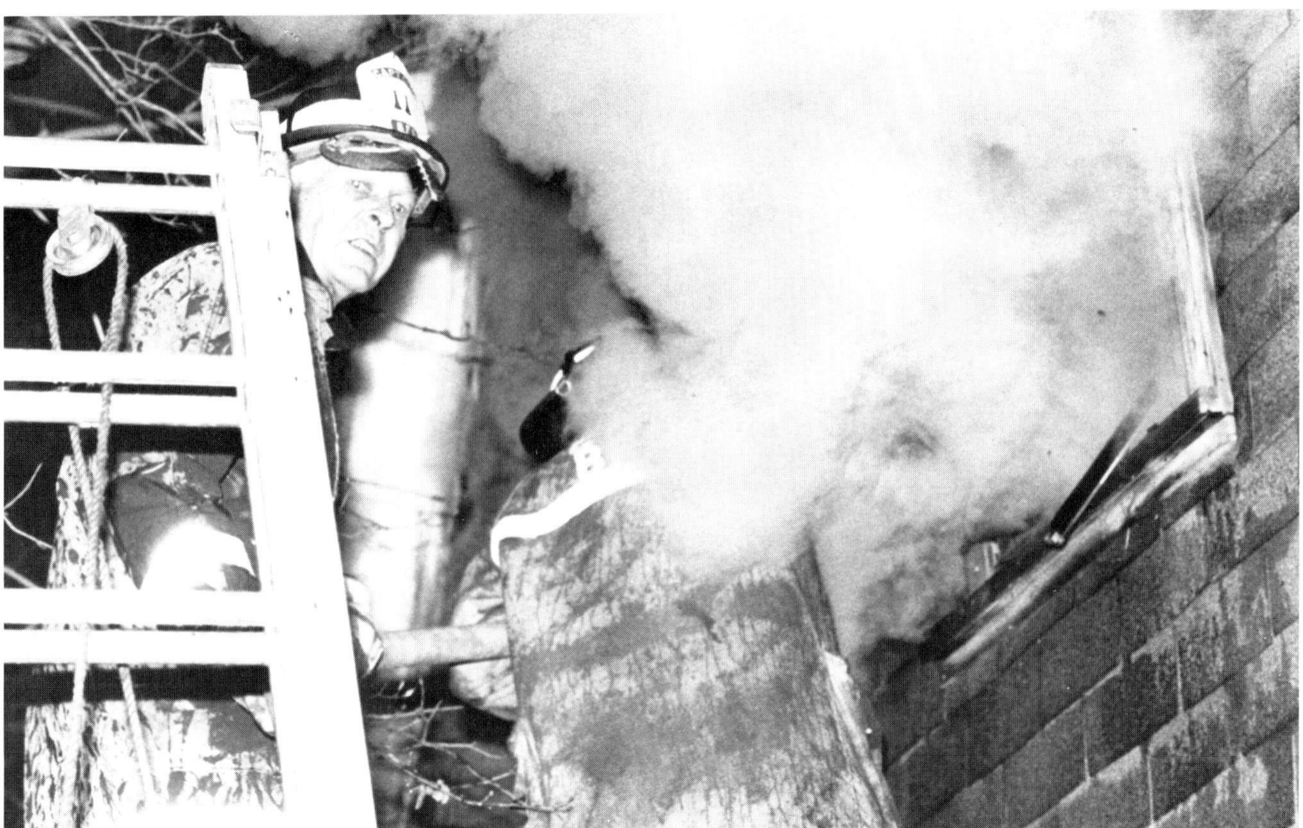

Braintree, 2nd Alarm, House

24
KEEP BACK 300 FEET.

Boston, 9 Alarms, Vacant Olive Oil Factory

Hull, 3 Alarms

Quincy, 4 Alarms

Weymouth, 2nd Alarm

Quincy

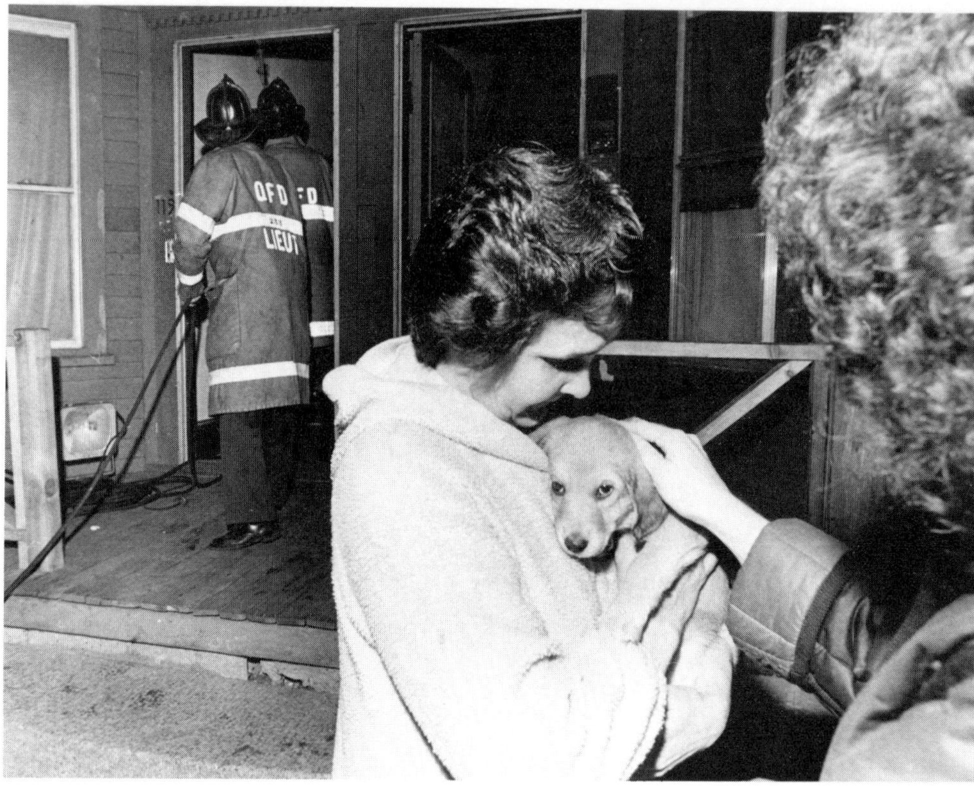

Quincy, 2nd Alarm, House

Braintree, Working Fire, House

Quincy, Working Fire, House

Quincy, 3rd Alarm

Quincy, 3rd Alarm

Quincy, 3rd Alarm

Brockton, 3rd Alarm, School

Quincy, 2nd Alarm

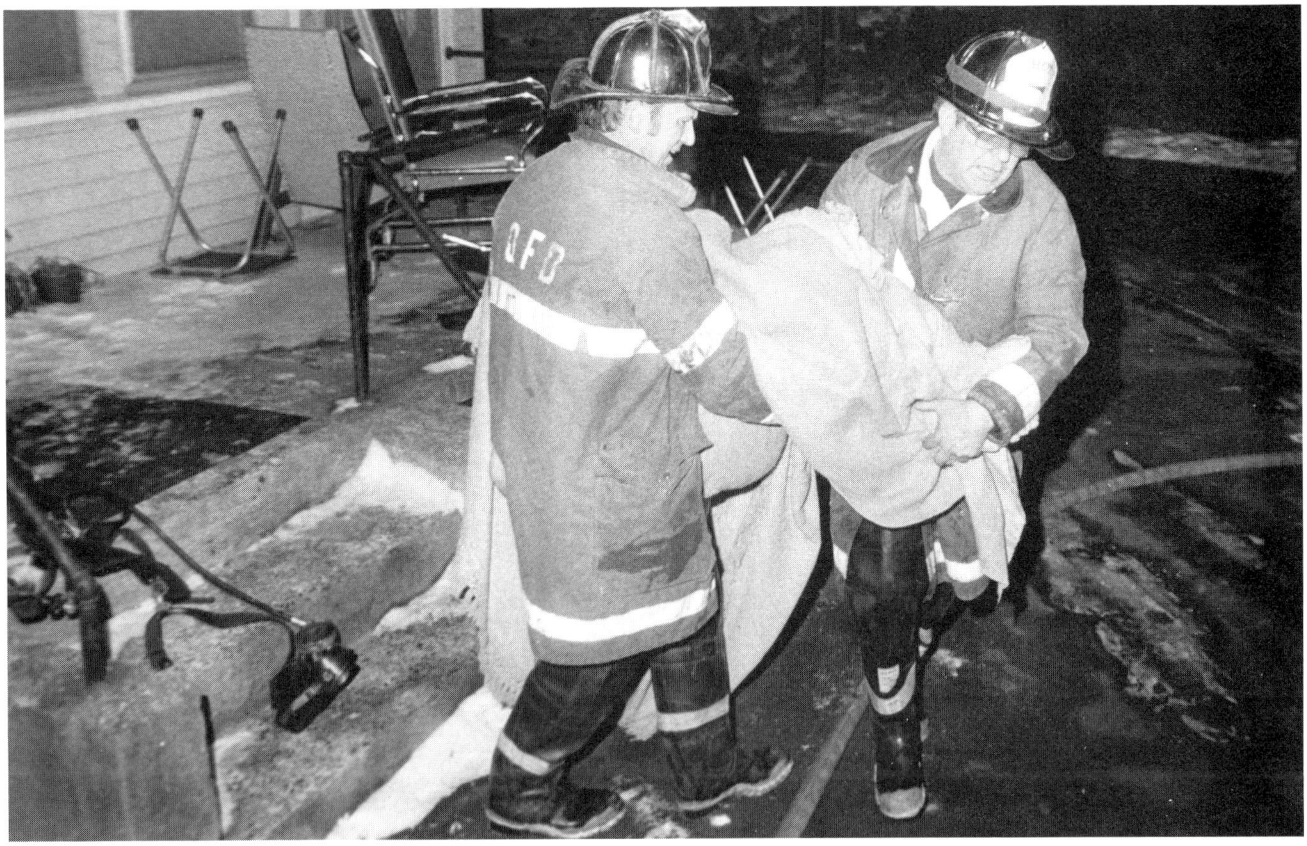

Quincy, 3rd Alarm, Nursing Home

Quincy, 3rd Alarm, Nursing Home

Braintree, Chemicals in a chemical recycling plant begin to explode. Three firefighters inside the plant were thrown out of the building from the explosions.

Braintree, Chemicals in a chemical recycling plant begin to explode. Three firefighters inside the plant were thrown out of the building from the explosions.

Boston, 3rd Alarm

Weymouth, 3rd Alarm

Randolph, 2nd Alarm, House

Quincy, 4th Alarm

Quincy, 3rd Alarm, Wall Collapse

Quincy, 4th Alarm

Quincy, 3rd Alarm

Quincy, 4th Alarm

Weymouth, Working Fire

Quincy, 2nd Alarm

Quincy, 2nd Alarm, School

Quincy, 2nd Alarm

Quincy, 2nd Alarm, Fatal

Quincy, 2nd Alarm, House

Hull, 3rd Alarm, Nantasket Beach

Boston, 5 Alarms

Braintree, Working Fire

Weymouth, 3rd Alarm

Quincy, 2nd Alarm, Restaurant

Boston

Brockton, Working Fire, House

Randolph, General Alarm

Boston

Weymouth, 3rd Alarm, Fleeing Wall Collapse

Quincy, 3rd Alarm, Restaurant

Weymouth, 3rd Alarm

Hingham, Working Fire

Whitman, General Alarm

Quincy, 3rd Alarm

Quincy, 4th Alarm

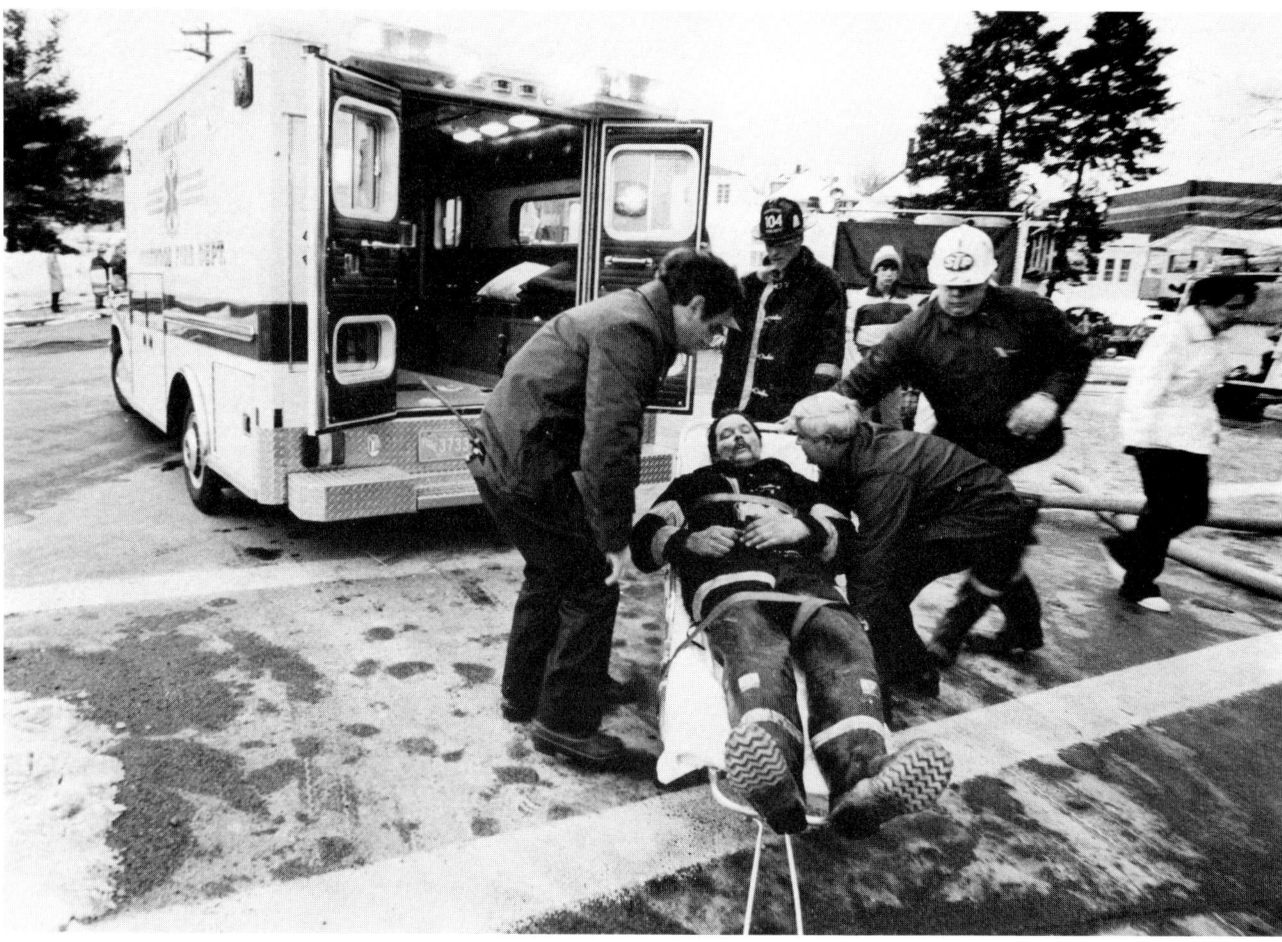

Norwood, 2nd Alarm

Hanover, General Alarm

Weymouth, 3rd Alarm

Pembroke, General Alarm

Quincy, 4th Alarm

Marshfield, Burst Hose

Quincy, 2nd Alarm, House

Weymouth, 2nd Alarm, House

Weymouth, 2nd Alarm

Weymouth, 2nd Alarm

Stoughton, Working Fire, Block of Stores

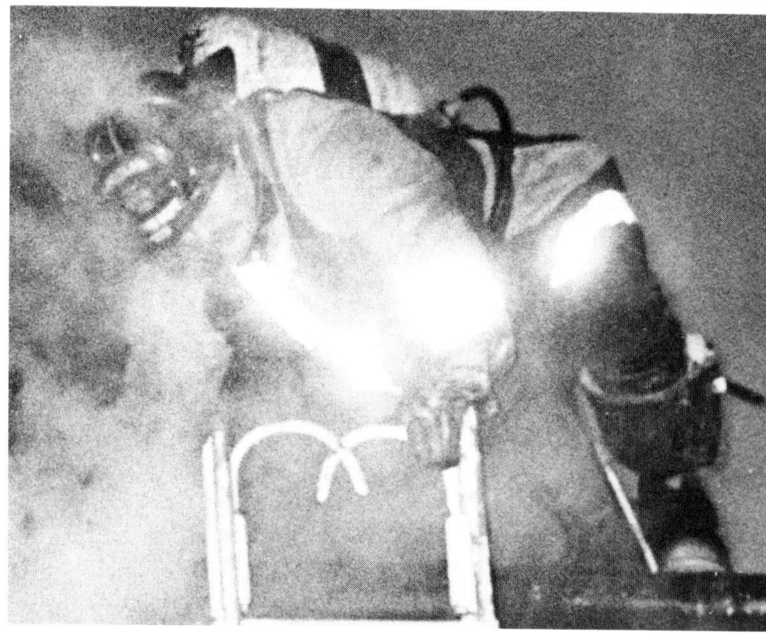

Braintree, 3rd Alarm, Block of Stores

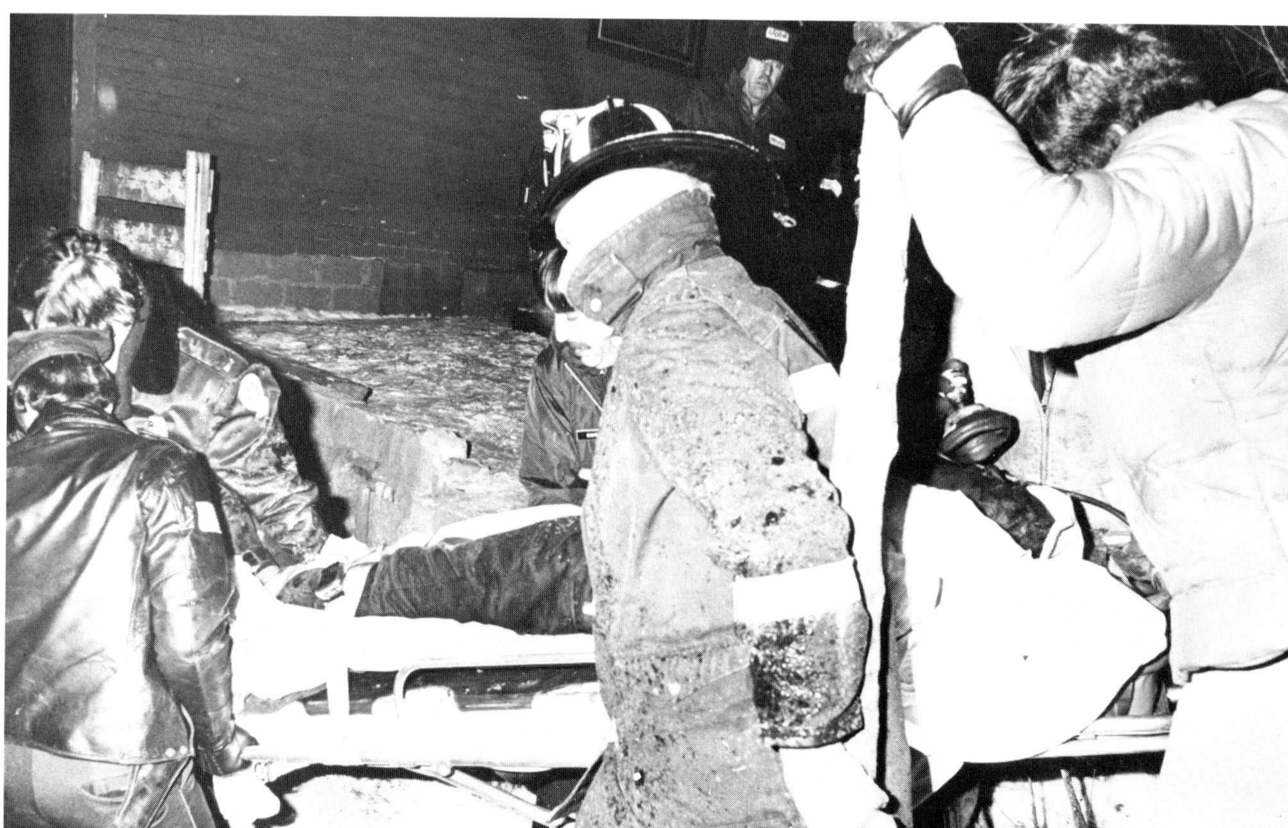

Weymouth, 2nd Alarm, House

Braintree, 3rd Alarm

Quincy, 2nd Alarm, Block of Stores

Fall River, Conflagration Destroyed 35 Homes, a Church and Several Stores.

Norwood, 3 Alarms

Quincy, 4th Alarm

Hanover, General Alarm, Factory

Holbrook

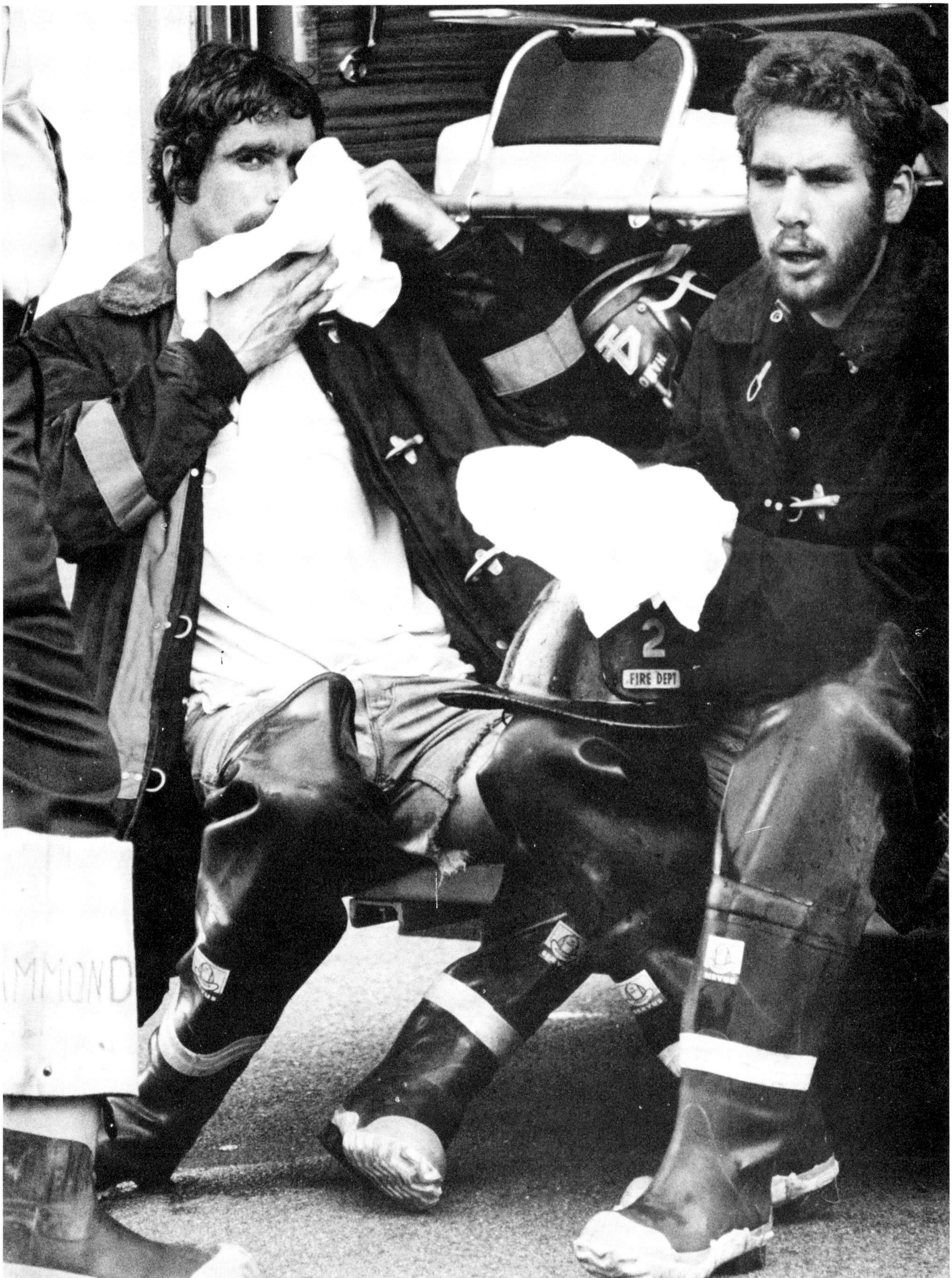

Hanover

Braintree. Boy drifted into a river on a raft, was rescued by firefighters.

Boston, Building collapse took the life of a firefighter at this 9 alarm fire.

Braintree, 37 tanks fell off a truck and exploded.

Providence, R.I., 4 alarms at the Outlet department store.

Hingham, Tour of Fire Station

Scituate, Muster

Hingham, Dispatcher

Hingham

Quincy, Working Fire, House

Quincy, 4th Alarm

Quincy, Working Fire

Weymouth, 3rd Alarm

Quincy, 4th Alarm

Weymouth, 2nd Alarm

Randolph, 2nd Alarm

Quincy, 2nd Alarm, Vacant Home

Quincy, 2nd Alarm, House

Boston, 9 Alarms

Quincy, 4th Alarm

Hanover, General Alarm

Quincy, 3rd Alarm

Rockland, General Alarm, Block of Stores

Rockland, General Alarm, Block of Stores

Quincy, 4th Alarm

Pembroke, General Alarm

Holbrook, Saving sign as fire destroys historic school

Milton, 3rd Alarm, House

Quincy, Truck Fire

Boston

5th Alarm, Boston

Weymouth

Weymouth

V.F.W. Fire, Braintree

Quincy, Ladder One, Setting up ladder pipe

Logan Airport, Boston

Logan Airport, Boston

Stoughton

Stoughton

Rockland

Weymouth, Lumber Yard Fire

Weymouth

Quincy

Sandwich

Sandwich

Sandwich

Sandwich

Weymouth

Quincy, Engine "B"

Braintree

Somerville

Dedham

Boston, W. 9th Street

Weymouth

Hanover, Factory

Fall River, Conflagration

Weymouth, Barn

Pembroke, Lumber Yard

Boston, W. 9th Street

Quincy, Engine "5"

Quincy, Aerial Tower 1

Quincy Square, 4th Alarm

South Boston, Tower 1 at 9th Alarm

Weymouth, General Alarm

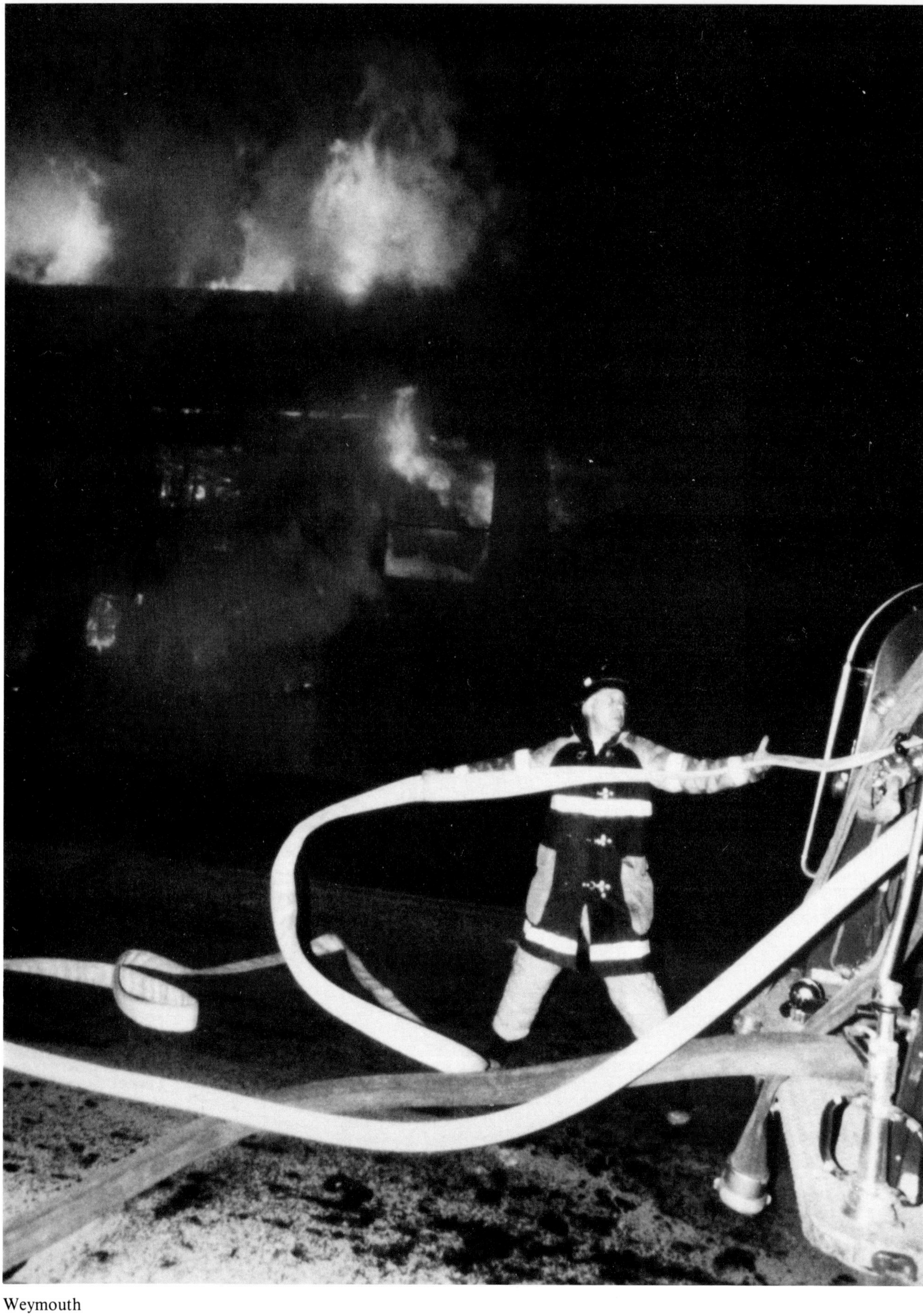

Weymouth

Stoughton

Weymouth

Brockton

Quincy, Quincy Lumber

Quincy

Weymouth

Quincy, 2nd Alarm, Factory

Quincy, Haz-mat Suits

Braintree, Truck Accident

Quincy, Video Taping Fire Scene

Braintree, Train Fire

Beverly, Train Accident

Braintree, Using Jaws of Life

Hanson, Truck Accident

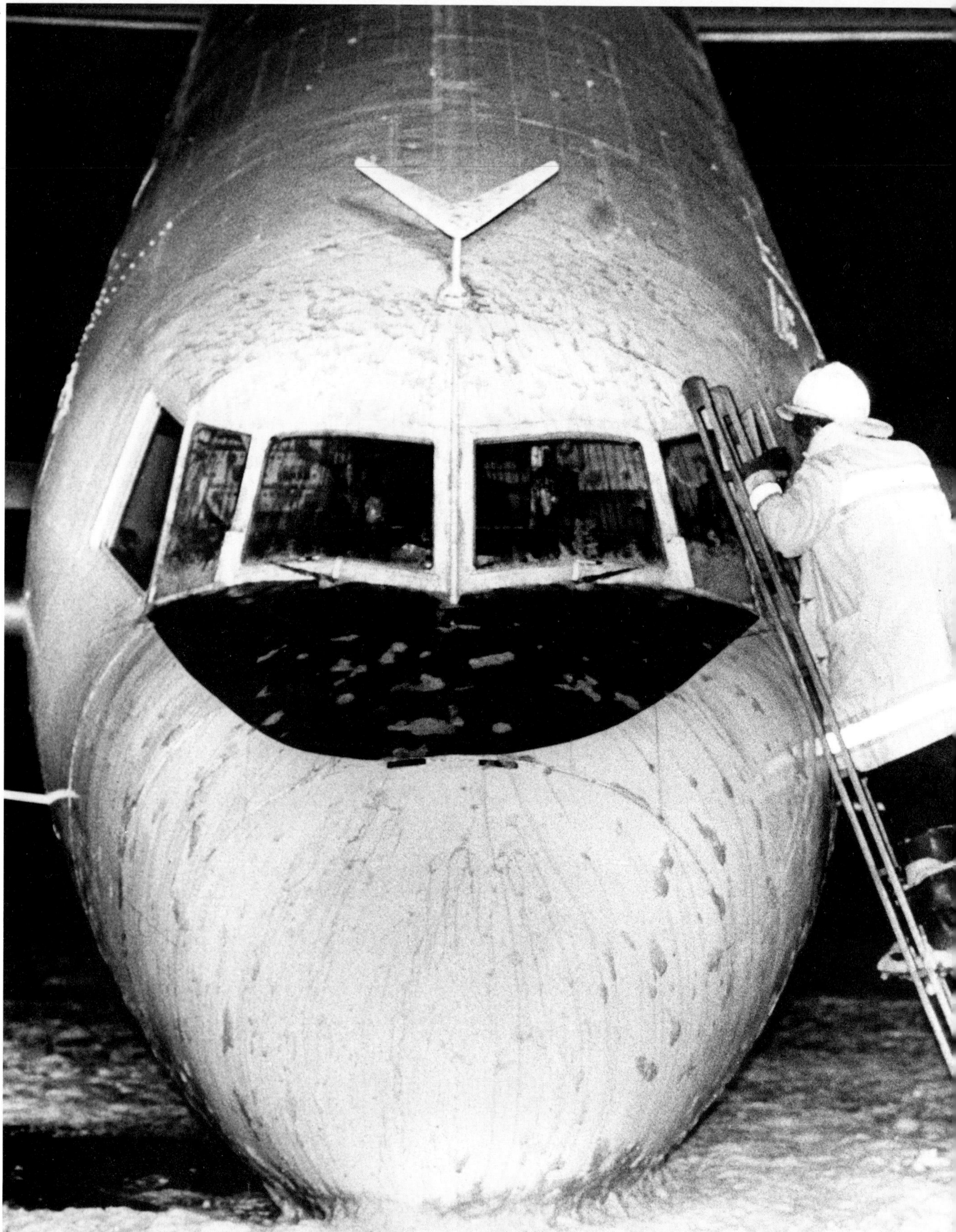

Boston, Logan Airport

Boston, 5th Alarm, Pier Fire

COAST GUARD 65608
FIREFIGHTER

Cohasset, Truck Accident

Cohasset, Foam Used at Truck Accident

Milton, Truck Accident

Braintree

Marshfield

Quincy

Quincy, Tree Down

Quincy, 4th Alarm

Randolph, 2nd Alarm, House

Quincy, 2nd Alarm

Weymouth, 2nd Alarm, Store

Weymouth, 2nd Alarm, Block of Stores

Norwood, Working Fire, House

Braintree, 3rd Alarm

Hingham, Working Fire, Stores & Apartments

Quincy, 2nd Alarm

Rockland, General Alarm

Boston, 5th Alarm

Quincy, 3rd Alarm

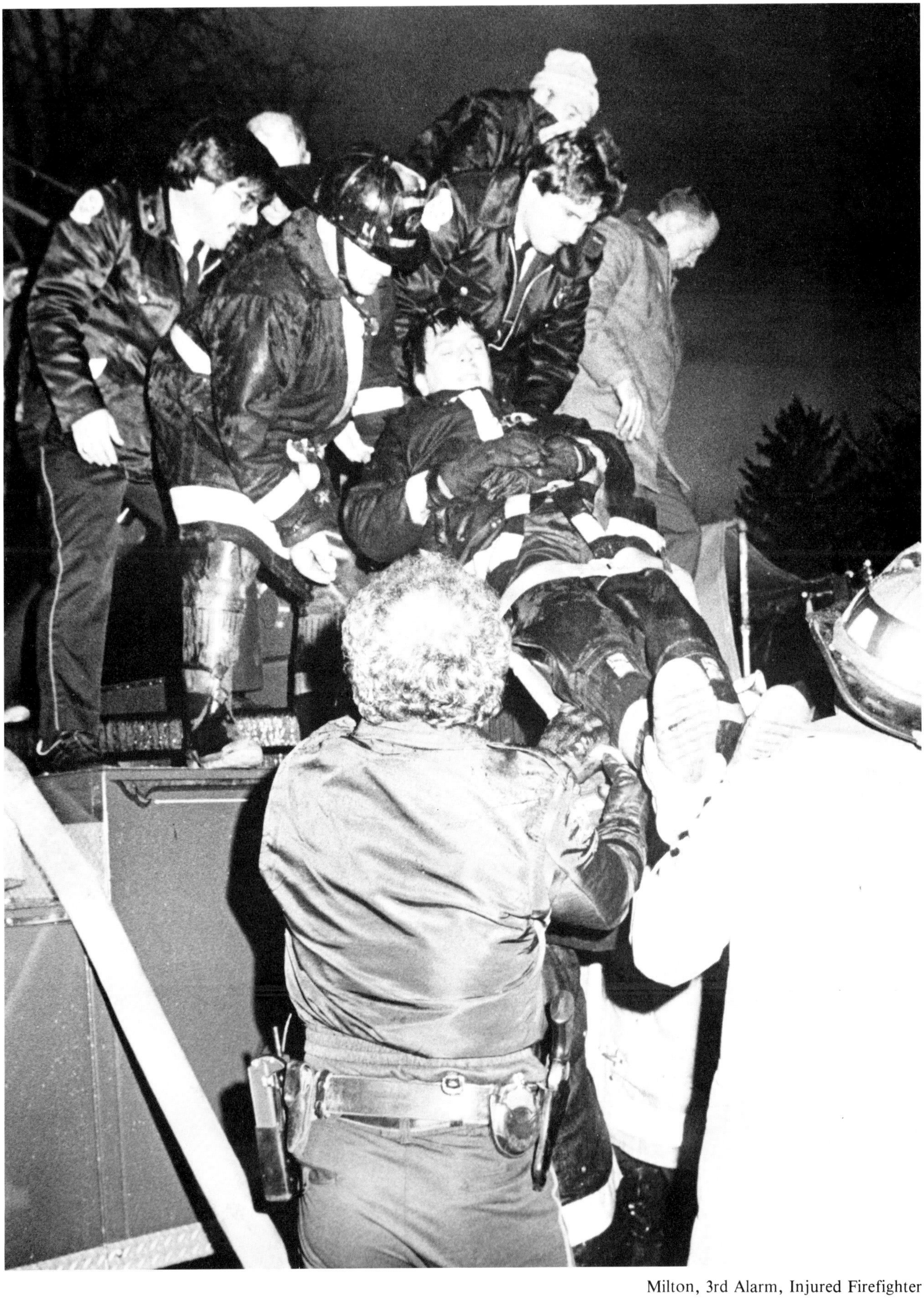

Milton, 3rd Alarm, Injured Firefighter

Hull, Vacant Hotel

Boston, 5th Alarm

Hull, 3rd Alarm

Quincy, 3rd Alarm, Restaurant

Whitman

Hanover, Muster

Milton, Posing for picture after filming movie.

Hanson, Muster Team

Dedham, 5th Alarm, Factory

Quincy, 3rd Alarm, Restaurant

Dedham, Brush fire from 5th alarm factory fire

Quincy, 4th Alarm

Duxbury, Working Fire

Rockland, General Alarm, Sandpaper factory

Weymouth, 2nd Alarm, Block of Stores

Hull, 2nd Alarm, Apartment Building

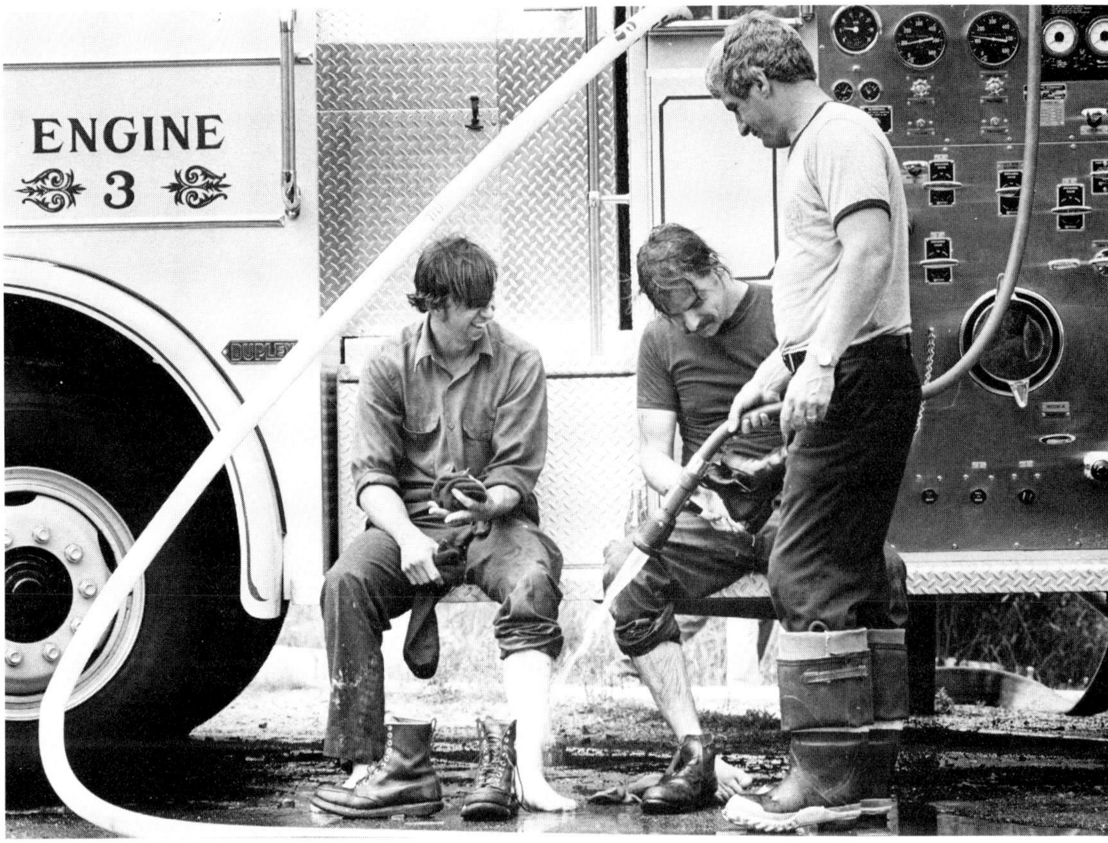

Cohasset

Weymouth, 2nd Alarm, Block of Stores

Kingston, Saw Mill

Quincy

Kingston

Quincy

KEEP BACK 300 FEET

Randolph, 4th Alarm, Moving Company

Hingham, Working Fire, House

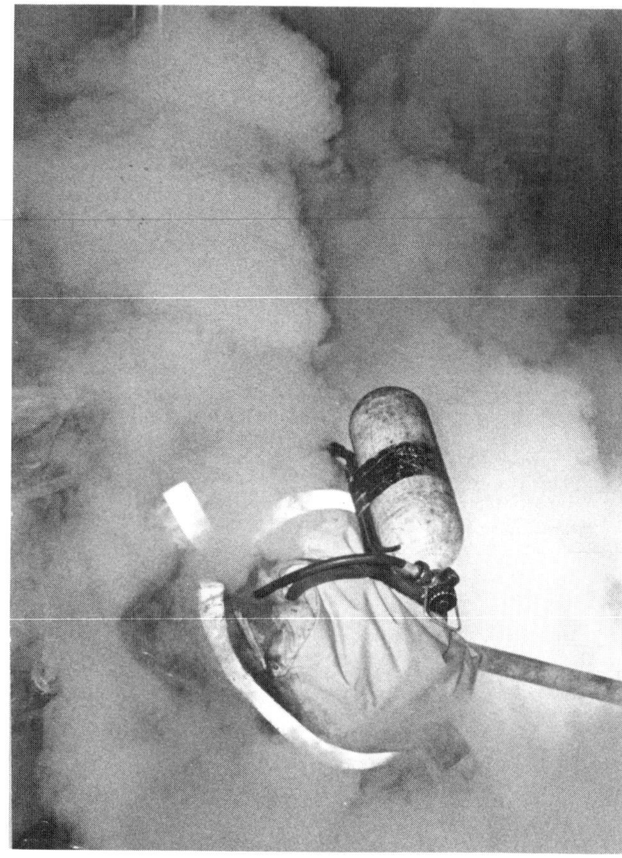

Quincy, 2nd Alarm, Block of Stores

Hull, 2nd Alarm, House

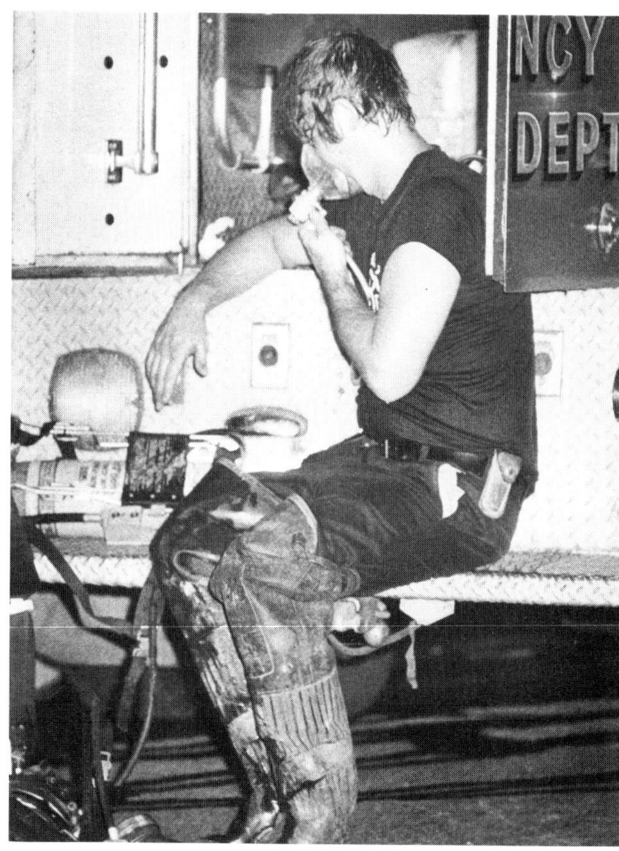

Quincy, 2nd Alarm

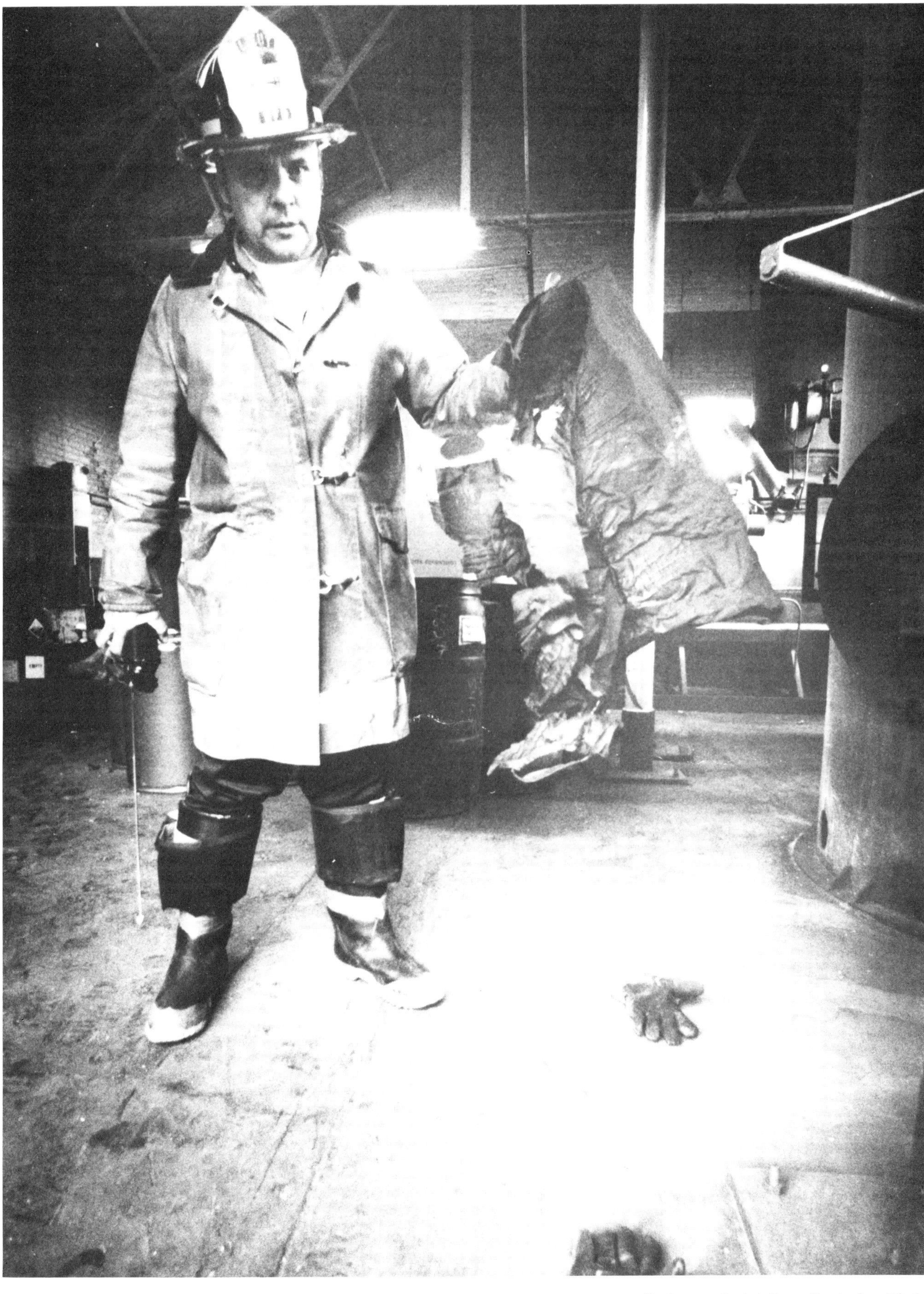

Braintree, Jacket from Explosion Victim

Quincy, 2nd Alarm

Cohasset

Quincy, Marsh Fire

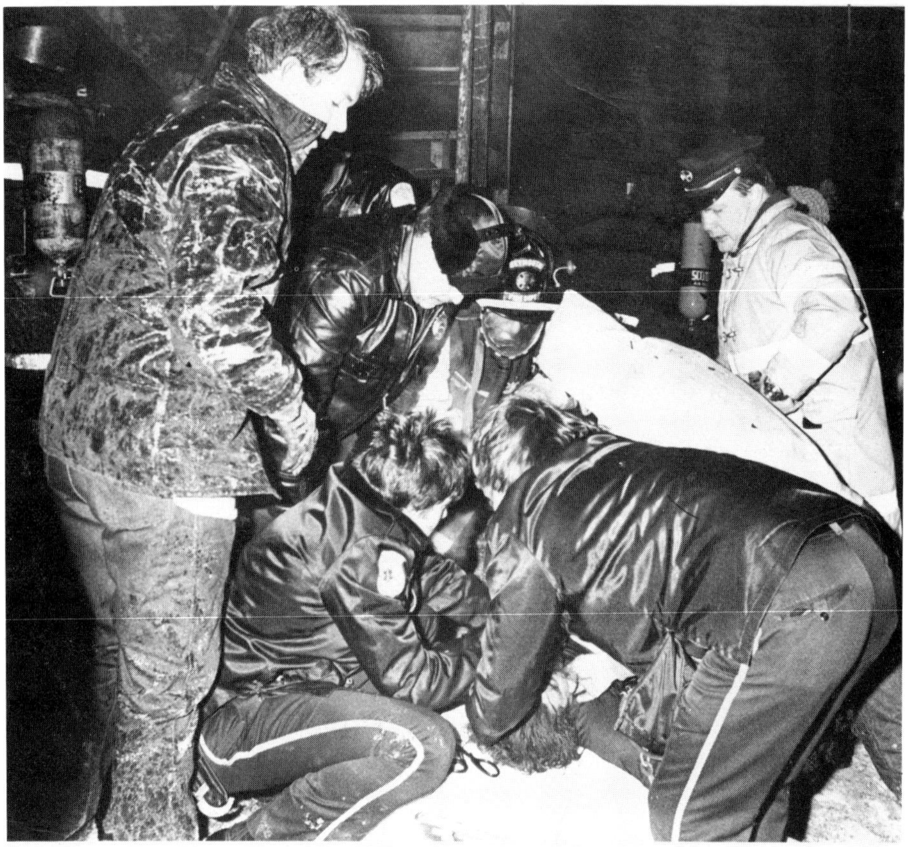

Weymouth, 2nd Alarm

Randolph, Searching for Victims of House Explosion

Dedham, 4th Alarm

Scituate, 3 Houses Involved

Milton, 3rd Alarm

Braintree, Working Fire, School

Boston, 9 Alarms

Boston, 8 Alarms

Boston, 3 Alarms

Quincy, 4th Alarm

Quincy, 4th Alarm

Weymouth, 3rd Alarm

Holbrook

Weymouth, 3rd Alarm

Whitman, General Alarm, Famous Toll House Cookie Co.

Fall River

Rockland

Braintree

Boston, 5th Alarm

Boston, 5th Alarm

Quincy, 4th Alarm

Quincy, 2nd Alarm, Block of Stores

Quincy, 2nd Alarm, House

Quincy, 3rd Alarm, Nursing Home

Quincy, 2nd Alarm

Brockton, 3rd Alarm

Boston, 5 Alarms, Church

Hingham, Working Fire, Industrial Park

Brockton, Working Fire, House

Marshfield, Church Fire

Kingston

Quincy, 2nd Alarm, House

Norwood, 3rd Alarm, Block of Stores

Braintree, Working Fire, House

Brockton, 3rd Alarm

Braintree, Barn

Weymouth, Barn

Randolph, 2nd Alarm, House

Quincy. Rapid Spread of Fire at 4th Alarm

QUINCY
LADDER
1

QUINCY
LADDER
1

Hanson

Norwood

Hull

Funeral For Boston Firefighter Killed in the Line of Duty.